Animal Groups

Insects

By Dalton Rains

www.littlebluehousebooks.com

Little Blue House is distributed by North Star Editions:
sales@northstareditions.com | 888-417-0195

Produced for Little Blue House by Red Line Editorial.

Photographs ©: Shutterstock Images, cover, 4, 7, 9, 11, 12–13, 15, 17, 19, 21, 22–23, 24 (top left), 24 (top right), 24 (bottom left), 24 (bottom right)

Library of Congress Control Number: 2022920061

ISBN
978-1-64619-809-2 (hardcover)
978-1-64619-838-2 (paperback)
978-1-64619-895-5 (ebook pdf)
978-1-64619-867-2 (hosted ebook)

Printed in the United States of America
Mankato, MN
082023

About the Author

Dalton Rains writes and edits nonfiction children's books. He lives in Minnesota.

Table of Contents

ant

Insects

An ant is an insect.

A butterfly is an insect.

butterfly

A ladybug is an insect.

ladybug

A dragonfly is an insect.

dragonfly

A beetle is an insect.

beetle

A fly is an insect.

fly

A bee is an insect.

bee

A firefly is an insect.

firefly

A cricket is an insect.

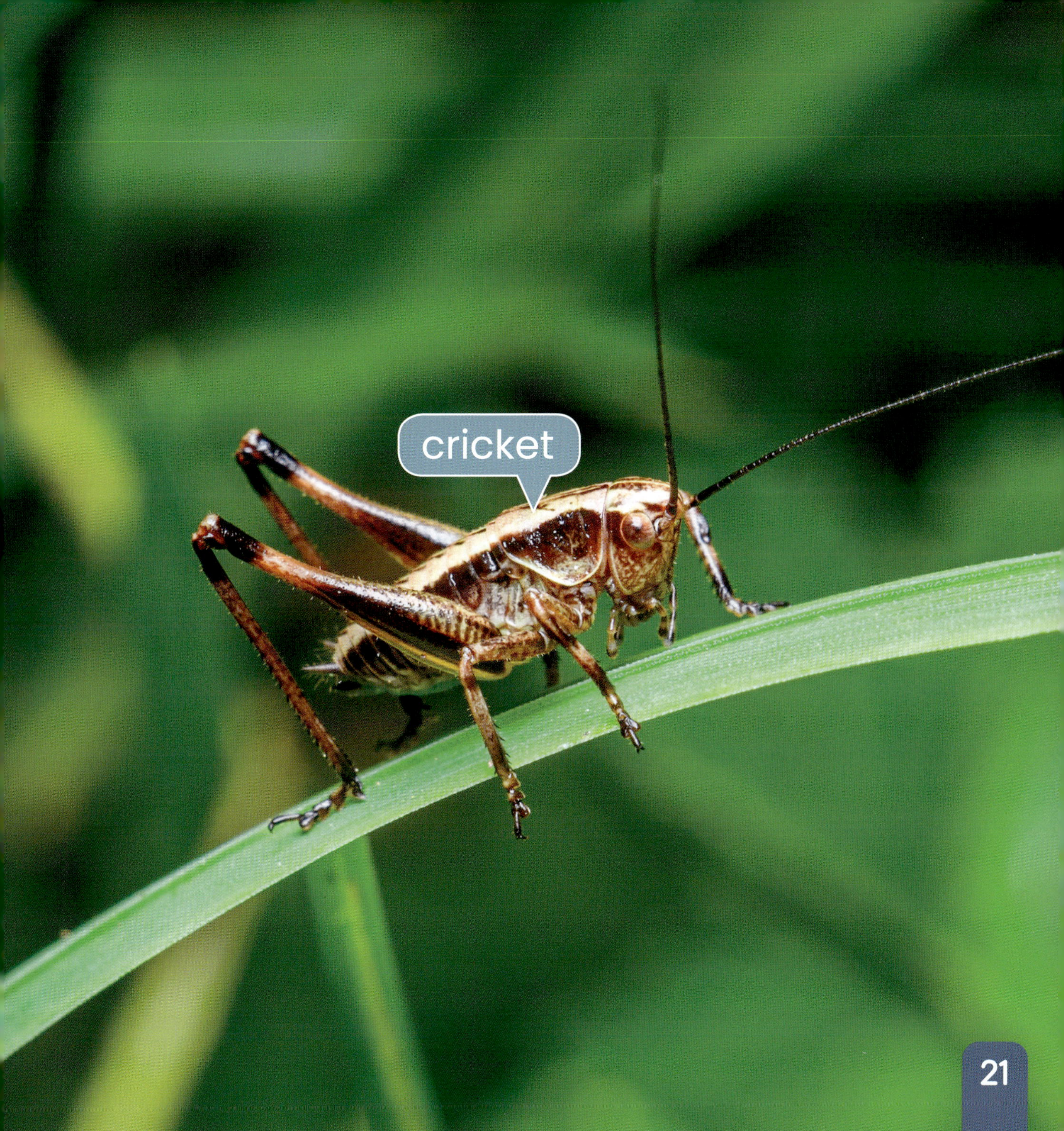
cricket

Is It an Insect?

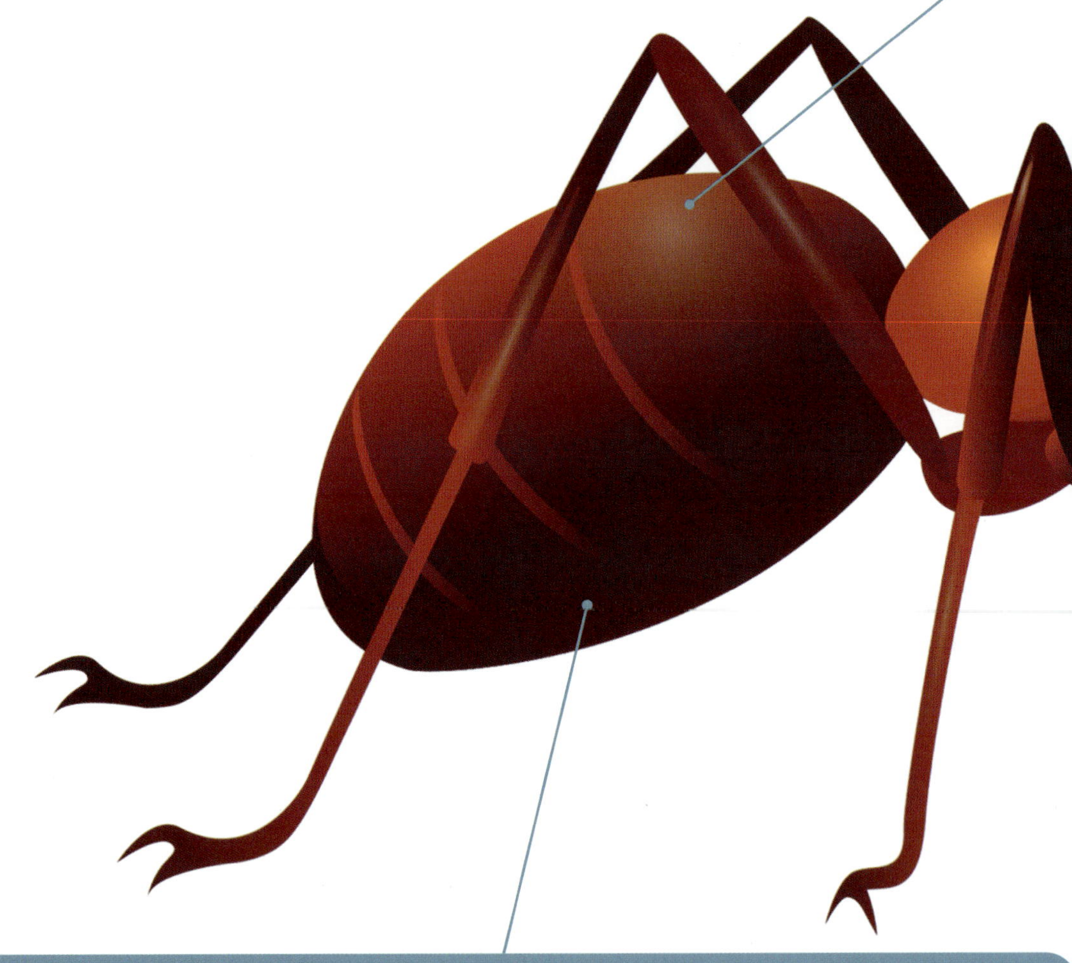

All insects have skeletons on the outside.

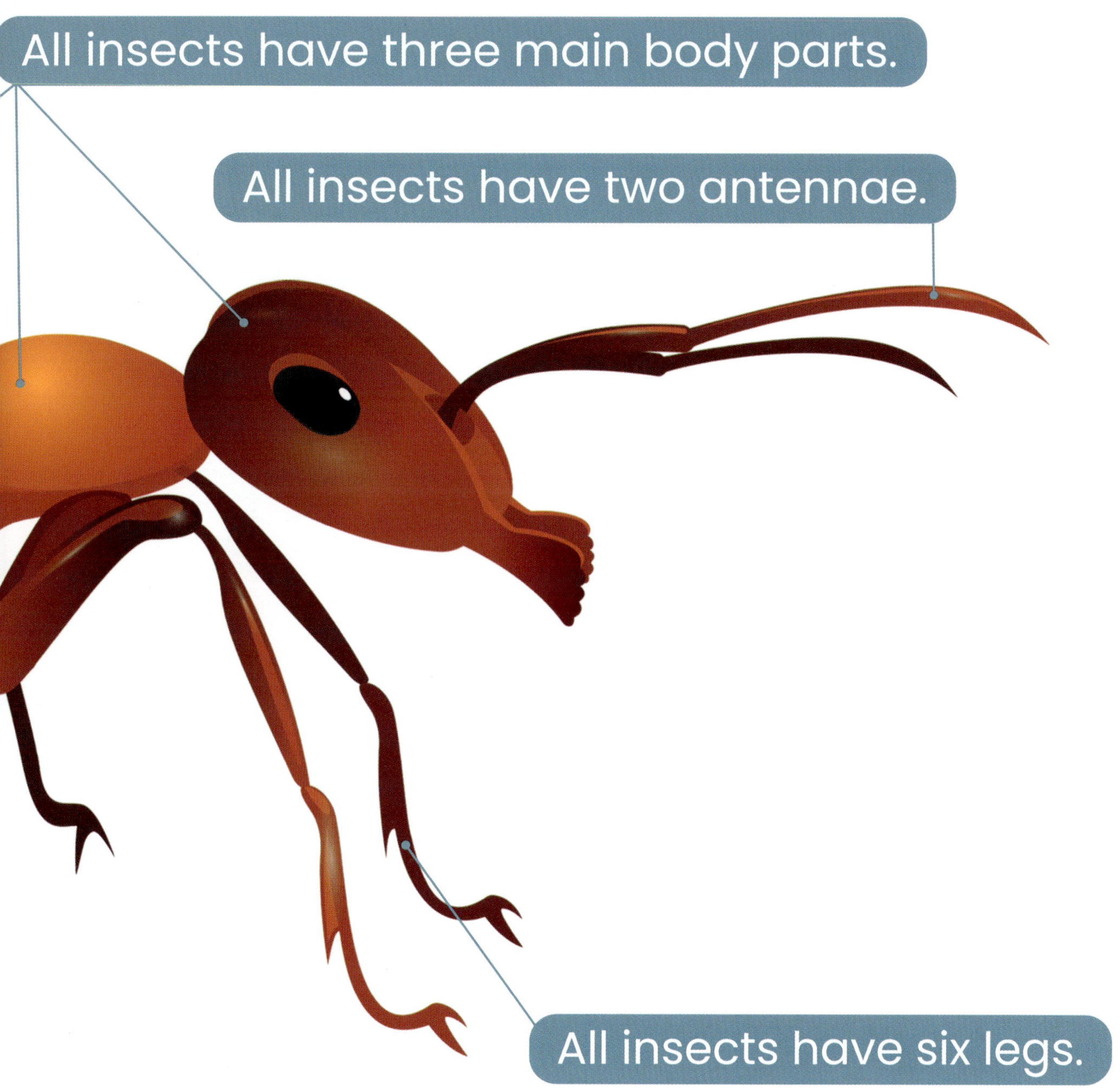
All insects have three main body parts.
All insects have two antennae.
All insects have six legs.

Glossary

butterfly

firefly

dragonfly

ladybug

Index